MECHANICAL DRAWING

BY

CHARLES F. JACKSON

Director of Drawing in the High Schools and Schools of College Preparation of Philadelphia; The Episcopal
Academy; Blight's School; Rittenhouse Academy; North Broad Street Academy; Forsythe
School; Martin's Academy; St. Luke's Academy, Bustleton; Media Academy;
The Philadelphia Central and the Germantown Branches Y.M.C.A.
Educational Classes; and Bethany College

PHILADELPHIA
J. B. LIPPINCOTT COMPANY
1896

PREFACE.

THE following work has been prepared for the use of students in Mechanical and Architectural Drawing, and is the result of teaching these branches for ten years. The subject of Projections is treated fully, beginning with the simple problems and advancing step by step in regular order to those more complex. The course is complete in one volume.

When principles have once been explained, they should be carefully studied until fully comprehended, as, with a few exceptions, they will not be referred to again, and after each drawing is made the student should make another with different data. This will impress more fully and with a greater degree of permanency the principles involved. Difficult terms are avoided as far as possible, and the explanations are so carefully and concisely made that any one of ordinary intelligence may easily comprehend them.

PREFACE.

A careful study of these problems is also recommended to Artists and Art Students, as they would undoubtedly obtain thereby a better comprehension of intersecting solids; and, furthermore, it is a most excellent preparation for the study of Perspective. It will be found that most young artists experience difficulty in working out problems in Perspective because of their lack of knowledge of the subject of Projections.

The accompanying plates are merely intended as sketches for guidance, no pretension being made in the direction of fine line work, which might discourage the student.

Those drawings in which new principles are involved are lettered so that the teacher may easily explain them to the entire class by referring to the letters, without fearing that the student will confound the lines or points referred to.

CONTENTS.

CONTENTS.

MECHANICAL DRAWING.

Instruments.

THE student should provide himself with the following articles:

Drawing-Board, T-Square, one 45° Triangle, one 30° Triangle, four Thumb Tacks, 6 H Drawing-Pencil, Scale, Irregular Curve, Rubber, Writing-Pen, Drawing-Paper, one bottle each of Black, Red, and Blue Drawing-Ink, a set of instruments containing at least one pair of Compasses, Dividers, Ruling-Pen, and Bow-Pen.

The lead in Pencil and Compasses should be sharpened upon sand-paper to a slender wedge-shaped point.

The Drawing-Board, T-Square, and Triangles should always be dusted previous to beginning work, and the instruments should be invariably wiped after using. Never leave ink to dry in Ruling-Pen.

To do good work, purchase the best instruments you can afford; you will never regret it.

5

Directions for Beginners.

Locate the drawing so that one side will not be nearer the adjoining edge of the paper than the opposite side is to its adjoining edge.   Lay off dimensions with the scale, placing it against the line ; not in space.

Always see that the head of the T-Square is firmly set against the Board, as many a mistake occurs with young students who do not regard this point.

Whatman's paper is the best,—hot-pressed being better for fine line, but cold-pressed permits of more erasures.   There are many other papers much cheaper which answer the purpose.

Avoid *digging* many deep holes in the paper with the points of Compasses or Dividers.

Hold Compass and Ruling-Pen firmly but *lightly*, and in inking avoid going over a line the second time, unless it be a shade line.   In ruling with pen do not place the edge of Triangle immediately against the line, but about one-sixty-fourth inch away, otherwise the Triangle may come in contact with the ink and cause it to spread.

To learn how to letter no better exercise could be recommended than copying ordinary text from any book, making the letters with all their details as near like the copy as possible.

Never do any work in a slipshod manner.

Definitions.

A solid has three dimensions, length, breadth, and thickness.

Parallel lines are everywhere equally distant from each other.

An angle is the difference in direction of two lines. The point of meeting of these lines or sides, produced if necessary, is the vertex of the angle.

A right angle is one whose two sides make an angle of 90° with each other.

An acute angle is less than a right angle.

An obtuse angle is more than a right angle.

A quadrant is an arc of 90°.

A triangle is a plane figure of three sides.

A quadrilateral is a plane figure of four sides.

A pentagon is a plane figure of five sides.

A hexagon is a plane figure of six sides.

A heptagon is a plane figure of seven sides.

An octagon is a plane figure of eight sides.

A nonagon is a plane figure of nine sides.

A decagon is a plane figure of ten sides.

An undecagon is a plane figure of eleven sides.

A dodecagon is a plane figure of twelve sides.

A circle is a plane figure of an infinite number of sides, or a plane figure bounded by a curved line everywhere equidistant from its centre.

An equilateral triangle has all its sides and angles equal.

An isosceles triangle has two sides and two angles equal.

A scalene triangle has all its sides and angles unequal.

A trapezium is a quadrilateral having no two sides parallel.

A trapezoid is a quadrilateral having only two of its sides parallel.

A parallelogram is a quadrilateral having its opposite sides parallel. Parallelograms are named as follows:

A square is a parallelogram having all its sides equal and its angles right angles.

A rectangle is a parallelogram having its opposite sides equal and its angles right angles.

A rhombus or rhomb is a parallelogram having all its sides equal, but the angles are not right angles.

A rhomboid is a parallelogram having its opposite sides equal, but its angles not right angles.

A polygon which has all its vertices on the circumference of a circle is said to be inscribed in the circle. The circle is circumscribed about the polygon.

A polygon which has all its sides tangent to a circle is said to be circumscribed about the circle. The circle is inscribed in the polygon.

A polyhedron is a solid bounded entirely by planes.

A prism is a polyhedron having two of its faces, called its ends or bases, parallel and the rest parallelograms.

A pyramid is a polyhedron having a polygon for its base, and for its sides it has triangles which have a common vertex, and the sides of the polygon for their bases.

The common vertex of the triangles is called the vertex, or apex, of the pyramid.

The axis of a prism is a straight line joining the centres of its ends ; and the axis of a pyramid is the straight line from its vertex to the centre of its base.

A cone is similar to a pyramid, except that its base is a circle.

PROBLEMS IN GEOMETRY.

Plate 1.

Fig. 1.—To draw a perpendicular to a given straight line, ab, from a given outside point, c.

From c describe an arc intersecting the line at a and b. From the points a and b describe arcs of any radius to d. From their intersection draw a line to c, the required perpendicular.

Fig. 2.—To bisect a given angle, acb.

From c describe an arc of any radius, in this case intersecting the straight lines at a and b. From a and b describe any arcs of equal radius intersecting at d. Join dc, which will bisect the angle.

10

Fig. 3.—To bisect a straight line or an arc of a circle.

From *a* and *b*, the extremities of both the straight line and arc, describe arcs of greater radius than one-half of *a b*. Join the intersecting points of these arcs *c d*, which is the line required.

Fig. 4.—To draw a perpendicular to a line *a b*, from a given outside point *c*, over its end.

Assume any point on *a b*, as *a*, and join with *c*. Bisect *a c*. From *d*, with radius *d c* describe the arc *c b a*. Join *c b*, the perpendicular required.

Fig. 5.—To trisect a right angle.

Let *a c b* be the right angle. From *c* describe an arc of any radius, intersecting the lines at *a* and *b*. With the same radius describe arcs from *b* and *a*, intersecting the quadrant at *d* and *e*. Join *c d* and *c e*.

Fig. 6.—To bisect an oblique straight line by the use of any right-angle triangle at the point *f*.

Place the hypothenuse of the triangle under and against the given line *a b*. Place any straight edge against the side *c d*. Now move the triangle, resting its base on the straight edge *c d*, the hypothenuse touching *f*. Then *f g* will be the required perpendicular.

Fig. 7.—To divide a given straight line into any number of equal parts.

At any angle with *a b* draw an indefinite line *b* 5. Parallel to it draw another from *a* toward 1′. On *b* 5 lay off one less than the required number of parts; do the same from *a* to 1. Join those of the same number, thus dividing the line *a b* as required.

Fig. 8.—Upon a given base *a b* to construct a regular pentagon.

From *a* and *b*, with radius *a b* describe circles intersecting at *f* and *g*. From *g*, with same radius, draw another circle. Produce 1 3 to *e*, also 2 3 to *c*. From *c* and *e*, with radius *c b* describe arcs intersecting at *d*. Join *a b c d e*.

Fig. 9.—Within a given circle to inscribe a regular pentagon.

Draw any two diameters at right angles to each other, as *a c* and *b d*. Bisect *d k*, and with *e* as a centre describe an arc of radius *e c* to *f*. From *c*, with *c f* describe arc *f g*. Connect *c g*, which will be one of the sides.

Note.—For all practical work, there is no better method of dividing a circle into an equal number of parts than by stepping it off with the dividers. After a little experience the divisions can be made with but two or three trials, care being taken not to prick the paper too deeply while experimenting.

Plate 2.

Fig. 10.—To describe a circle through three points, or to circumscribe a circle about a triangle.

The points are *a b c*. Connecting them with straight lines will form the triangle. Bisect the lines *a b* and *b c*. From *d*, their intersection, with radius *d a* describe the circle.

Fig. 11.—To inscribe a square within a given triangle, *a b c*.

Draw *a d* perpendicular to *a b* and equal to it. From *c* draw *c j* perpendicular to *a b*. Connect *d* and *j*, intersecting *a c* in *h*. Draw *h e* perpendicular to *a b*. Describe arc *e h* to *f* and *f e* to *g*. Connect *f* and *g*.

Fig. 12.—To construct a regular octagon on a given base *a b*.

Erect perpendiculars at *a* and *b* and bisect the exterior right angles, making *b d* and *a c* each equal *a b*. Connect *c* and *d*. On *h* lay off the distance *f e* from *f*. Lay off the same on *g* from *e*. Draw through *g h*, producing the line indefinitely. Draw *d l* and *c k* parallel to *f h*. Make *h m* and *g k* each equal *l h*. Connect *a b d l m n k* and *c*.

Fig. 13.—To construct an equilateral triangle when the altitude $a\,b$ **is given.**

Draw $c\,d$ perpendicular to $a\,b$. With any radius describe a semicircle. With same radius describe arcs from c and d, intersecting the semicircle at g and h. From a draw indefinitely through g and h. Through b draw $e\,b\,f$ parallel to $c\,a\,d$.

Fig. 14.—On a given side $a\,b$ **to construct a regular hexagon.**

From a and b describe arcs of radius $a\,b$ intersecting at g. From g, with radius $g\,b$ describe a circle. With same radius from f lay off e, and d from c. Join $a\,b\,c\,d\,e$ and f.

Fig. 15.—On a given base $a\,b$ **to construct a regular polygon of any number of sides (say 7).**

From b with $a\,b$ describe a semicircle, and divide it into as many equal parts as there are sides in the polygon. Connect the second point 2 with b. Bisect $b\,a$ and b 2. From c, the intersection of the bisectors, describe a circle with radius $c\,a$. Lay off $b\,a$ seven times on the circle and adjoining points.

Fig. 16.—The base $a\,b$ **and the angle at the vertex** h **of an isosceles triangle being given, to draw the triangle.**

Produce the base indefinitely, say to c. With radius $a\,c$ describe a semicircle.

Make the angle *d a c* equal to *h*.   Bisect the angle *d a b*.   Make *f b a* equal *e a b*, and produce both to meet in *g*.

Fig. 17.—To draw a circle of a given radius *a b*, **tangent to a given straight line** *c d*, **and also to a given circle** *e*.

Draw *f k* parallel to *c d* at the distance *a b*.   With the radius *e g* equal to the sum of the radii of the two circles, describe an arc intersecting *f k* in *h*.   From *h*, with radius *a b* draw the circle required.

Fig. 18.—At a given point *b* **on a circle to draw a tangent.**

Draw any chord *a b* from *b* and bisect it.   From *b*, with radius *b e* describe arc *d f*.   From *e*, with radius *e d* describe arc cutting *d f* in *f*.   Connect *b* and *f*.

Plate 3.

Fig. 19.—To draw a circle through a given point *c* **and tangent to a given line** *a b*.

Erect a perpendicular to *a b* at *a*.   Connect *a c*.   Bisect *a c*, intersecting *a d* in *d*, the required centre.

Fig. 20.—To draw a circle tangent to a given circle *b* **at a given point** *a* **on the circle, and through a point** *c*.

Join *a b*, producing the line indefinitely. Connect *a c* and bisect it. The point *d*, at which the bisector and *b a* produced intersect, is the centre required.

Fig. 21.—To draw a circle of given radius *o* **tangent to two other given circles** *a* **and** *b*.

From *b* with the sum of the radii of the circle *b* and the required circle describe an arc. With the radius of the circle *a*, plus *o*, describe an arc from *a*, the two intersecting at *c*, the centre of the required circle.

Fig. 22.—To draw two circles tangent to two given circles *a* **and** *b* **at a point** *c* **in one of them.**

Draw through *b* and *c* indefinitely a straight line. From *c* at *f* and *d* lay off the radius of circle *a*. Connect *d* and *f* with *a*. Bisect *d a* and *f a* with perpendiculars, cutting *b c* produced in *e* and *g*, the required centres.

Fig. 23.—Within a triangle *a b c* **to inscribe a circle.**

Bisect any two of the angles. The point *d* at which the bisectors meet is the centre of the required circle, and a perpendicular from *d* to any of the sides is the radius.

Fig. 24.—To connect two given parallel straight lines $b\,a$ **and** $d\,c$ **with arcs of circles tangent to them at** a **and** d **and cutting the line** $a\,d$ **at any point as** g.

Erect perpendiculars at a and d. Bisect $a\,g$ and $g\,d$ by perpendiculars, intersecting the perpendiculars from a and d at f and e. Draw the arc $a\,g$ with radius $f\,a$ from f, and draw the arc $g\,d$ with $e\,g$ from e.

Fig. 25.—Within a square to draw four equal semicircles, their diameters forming a square and each tangent to two sides of the square.

Draw the diameters and diagonals. Bisect $e\,d$ and $c\,h$ with perpendiculars. Join $k\,i$, cutting $g\,h$ in s. Lay off the distance $r\,s$ from r on $u\,w$ and v. Connect s, u, w, v, and where these lines intersect the diagonals are the required centres.

Fig. 26.—To draw three equal circles within an equilateral triangle, each circle tangent to two others and to one side of the triangle.

Bisect the angles, and then bisect the angle $e\,b\,a$, cutting $f\,c$ in f. From m, with radius $m\,f$ describe a circle cutting $a\,d$ and $b\,e$ in g and h. From the centres $f\,h\,g$, with radius $h\,e$ describe the required circles.

2

Fig. 27.—To inscribe three equal circles within an equilateral triangle tangent to two sides and to two other circles.

Bisect the angles, cutting the sides in *d e* and *f.* With radius *f d* describe arcs from *d, e,* and *f* through *o m g* the centres required.

Plate 4.

Fig. 28.—To circumscribe any number of equal circles about a given circle, tangent to it and to each other.

Divide the given circle by twice the number of diameters as circles required. Draw a tangent *c a e* to the given circle at *a.* Lay off *e d* equal to *e b* and erect the perpendicular *d f.* Where *d f* cuts the diameter *b a* produced is the centre of one of the required circles.

Fig. 29.—To draw three equal circles within a given circle, tangent to it and to each other.

Divide the circle by diameters into six equal parts. Produce one of the diameters as *h f.* Lay off *f d* equal to *f h.* Connect *n d.* Bisect the angle *h d n.* With centre *h* and radius *h g* describe a circle. *m, o,* and *s* are the centres of the required circles.

Fig. 30.—To divide a circle into any number of parts which shall have an equal area.

Divide the diameter into twice the number of equal parts required; ten in this case, as the number required is five. Number them, and with 1 and 9 as centres and radius 9 o describe semicircles from each on opposite sides. Likewise from 2 and 8 with 8 o, from 7 and 3 with 7 o.

Fig. 31.—To divide a circle into concentric rings having equal areas.

Divide the radius *d e* into any required number of parts. On the diameter *d e* draw a semicircle and erect perpendiculars to *d e* at *h g* and *f*. Through the intersection of these lines with the semicircle *d e* describe concentric circles from the common centre *d*.

Fig. 32.—To draw the involute of a circle.

Divide a circle into any number of equal parts. Draw radii to these parts, and where these radii intersect the circle draw tangents, as at 1, 2, 3, 4, etc. On *a* from 2 lay off the distance 2, 1. On *b* from 3 lay off twice the distance from 1 to 2, on *c* from 4 three times the distance, etc. A curve drawn through these points *a*, *b*, *c*, *d*, *e*, and *f* is the involute of the circle, which is simply the curve generated by a point at the end of a string which is being unwound from the circle beginning at the point 1.

Fig. 33.—The spiral of Archimedes.

Divide a circle into any number of equal parts, drawing radii to them. Divide the radius *o a* into the same number of equal parts. From the centre *o* with radius *o* 1 describe an arc intersecting *b* at *m*. With *o* 2 from *o* intersect *c* at *n*. With *o* 3 cut *d* at *r*, etc. Draw the curve through *o m n r*, etc.

Fig. 34.—To draw the involute of a square, *a b c d*.

Produce all of the sides of the square. From *d*, with radius *d c* describe the arc *c f*. From *a*, with *a f* draw the arc *f e*. From *b*, with *b e* draw *e g*, etc.

Fig. 35.—The major and minor axis of an ellipse being given, to draw the ellipse.

In the figure, *a b* is the major axis, *c d* the minor axis, and *f f'* the foci. To obtain the foci, describe arcs of radius equalling one-half the major axis, upon the major axis from either extremity of the minor axis, as from *c* to *f* and *f*. Assume any point on the major axis as 5. With radius *a* 5 describe arcs from *f'* to 5' and 5", and from *f* to 5''' and 5''''. With the balance of the major axis as radius, as *b* 5, describe arcs from *f* and *f'*, intersecting those already drawn at 5' 5" 5''' and 5''''.

Assume another point on the major axis as 4 and repeat. Then with 3, etc. Draw the curve through the points thus found.

Fig. 36.—To draw an approximate ellipse.

Suppose the major and minor axes (as $a\,b$ and $d\,e$) are given. Lay off one-half the minor axis from b to f. Then $f\,a$ will be the radius $g\,k$ for the top and bottom. Lay off one-half of $f\,c$ from f toward b, which produces x, then $x\,b$ is the radius for the ends.

Plate 5.

Fig. 37.—Having the abscissa 1 2 and the ordinate 1 4 given, to draw a parabola.

Draw 3 4 parallel to 1 2, and 2 3 parallel to 1 4. Divide 1 4 and 3 4 into the same number of equal parts. Join $a'\,b'\,c'$, etc., with 1. Through $a\,b\,c$, etc., draw parallels to 1 2. The intersections of a and a', b and b', etc., are the points through which the curve passes. The half below 1 2 is made in the same manner.

Fig. 38.—Given the diameter 1 2, the abscissa 1 3, and the double ordinate 4 5, to draw an hyperbola.

Draw the rectangle of which 4 6 and 4 5 are two sides. Divide 4 6 and 3 4

into the same number of equal parts. Join 4, 7, 8, etc., with 2, and join 12, 13, 14, etc., with 1. The curve will travel through the points of intersection of 2 11 with 1 12, 2 10 with 1 13, etc. Duplicate this below 1 3, and on the opposite side through 17, 18, and 2.

Fig. 39.—A cycloid is a curve generated by a point on the circumference of a circle which rolls upon a straight line.

Let 9 4 be the diameter of the rolling circle, and 0″, 1″, 2″, 3″, 4″, equidistant points, be the centres of this circle in different positions. Upon the arc 1′ *a*, drawn from 1″ lay off 9 1′. Upon 2′ *b*, drawn from 2″ lay off twice the distance 9 1′. Upon 3′ *c*, lay off three times the distance, etc., these points *a b c d*, etc., being the points through which the cycloid is drawn. These divisions may be conveniently laid off upon the circle having the diameter *s m*, as *a′ b′ c′*, etc., and then projected, as from *a′* to *a*, *b′* to *b*, etc. The curve may be drawn by describing an arc from the centre 1′ with radius 1′ 9, joining 9 and *a*. Produce *a* 1′ and *b* 2′ and their intersection will be the centre for *a b*. Repeat this for *c d*, *d e*, etc.

Fig. 40.—An epicycloid is a curve generated by a point on the circumference of a circle rolling on the exterior of another circle.

Let the circle of which 1′ is the centre and 1′ *d* the radius roll upon the arc of a circle *a b*, called the base circle. From *x*, the centre of the base circle, with radius *x a* plus the radius 1′ *d*, describe the arc *c d*. Upon *a b* from the central vertical line lay off equidistant points 2, 3, 4, etc., joining them with *x* and producing them to the arc *c d*. From the intersection of these lines produced, with *c d* describe circles with radius 1′*d*, and from their point of tangency lay off upon each as many of the divisions 2, 3, 4, etc., as they are removed from the first circle, 1′. Thus, on the arc drawn from 2′, lay off one division from 2 ; on that drawn from 3′, lay off two divisions from the point of tangency 3, etc.

A hypocycloid is a curve generated by a point upon the circumference of a circle rolling upon the inside of another circle, and the points are found as in the epicycloid, *e f* being the arc upon which are located the centres of the rolling circle.

Fig. 41.—If the diameter of the rolling circle equals one-half that of the base circle inside of which it rolls, the hypocycloid becomes a straight line.

THE HELIX.

Fig. 42.—If a cord moving forward uniformly, be wound about an evenly revolving cylinder, it will form a helix, and the distance forward, travelled by the cord or helix in one revolution of the cylinder, is called the pitch.

The semicircle shows one-half of the plan of the cylinder. Divide this into any number of equal parts, say twelve, which in the completed circle would be twenty-four. Divide the pitch $a\,b$, into the same number of equal parts. Since the helix begins at the middle point a, in the top of the elevation, we will give the same number to the top line as that given to the radius in the plan from which the point is projected. Thus, number the plan 1, 2, 3, etc., and the elevation 1', 2', 3', etc., as shown in the plate. As the intersection of the projection of 1 and 1' gives one point in the helix, so will the intersection of the projections of 2 and 2' give the next. Do this with all of the points shown. Then 6 will also answer for 9, 5 for 10, 4 for 11, etc. Through these points of intersection draw the helix.

24

Fig. 43.—To draw the projection of a V-threaded screw, two and one-half inches diameter, four threads per inch.

Join *a b c* and *d*. Lay off one-quarter inch spaces upon *a b*. From *a* draw a line toward the middle, inclining downward at an angle of 30°. From 1 draw a line upward at an angle of 30° to meet the line from *a*. Then repeat, from 1 downward, 2 upward, etc. Project the intersection of the lines drawn from *a* and 1 to the line *c d*, which gives 3. Project the intersection of 1 and 2, which gives 4, etc., when the V's are drawn as on *a b*. Join *a* and 3 with a curved line obtained as in Fig. 42. Join 1 and 4, etc., and then also join the bases of thread.

It will be observed that the outside points on *c d* are opposite the inside points on *a b*.

Conventional threads are usually drawn without making the helical curves, but simply joining the proper points with straight lines. More frequently still, the V's are not drawn at all.

PROJECTIONS.

Plate 6.

Fig. 44.—To make a mechanical drawing of a rectangular prism six inches long, four inches high, and two inches thick.

The impossibility of showing all three dimensions in a single drawing is at once apparent. Hence, separate views of various sides are shown. In this case we have three, a front view or front elevation, side view or elevation, and the top view, or plan.

The top view, as its name indicates, is a view of the top, or in other words a view of the nearest side of the adjoining view, just as the front elevation is a view of the lower or nearest side of the top view. The end elevation is a view of the end of the front elevation.

To understand fully the theory of projections, let the student assume a position immediately in front of an ordinary dwelling-house. He has now a view of

26

the front, or a front elevation. Assume one end of the house to be attached to the front with hinges. Then, in imagination, swing this end around until it stands on a line with the front. Now, supposing the roof is flat, with one end hinged to the front, we also swing it forward on a line with the front and side, and have then the front, the end, and the roof all on one plane, in which they may be seen in their true size and shape.

The imaginary lines around which these sides have swung are called the axes of projection.

In drawing the figure (44), we have followed out exactly the method above described, only the different views are removed slightly from each other, and the axes of projection placed midway between them.

To draw the figure begin by assuming a point at *a*, drawing from this point, with pencil, a horizontal line with the T-square and a vertical line with the right-angle triangle, being sure that both these lines are long enough to include the other views. From *a*, toward the right lay off six inches ; and upward, on the vertical line lay off four inches, and through the latter draw a horizontal line, and from the former draw a vertical line, as in the first instance. This completes the front elevation.

On the line *a b*, which has already been drawn, assume the point *b*, which should be about as far from the upper edge of the sheet or space in which it is drawn as *a* is from the lower. Draw an indefinite horizontal line from *b*, and from that line toward *a* lay off two inches, through which draw another horizontal line, which completes the plan. Midway between these two views draw the line *f g*, which is one of the axes of projection ; *n o*, being the other, is as near to the right-hand edge of the front view as *f g* is to the top. Place the needle point of the compasses at *v*, the intersection of *f g*, and *n o*, and the pencil point at the intersection of the upper line of the top view with *n o*, and describe an arc to *f g*. With the needle point still at *v*, place the pencil where the lower line of the top view intersects *n o* and describe an arc to the line *f g*. This is simply transferring the thickness of the top view, as shown on the axis *n o*, to the axis *f g*, from which points vertical lines may be drawn through the top and bottom as projected from the front view, thus completing the end view.

The drawing is now pencilled and only requires inking to be completed. The lines of the object are black and should be drawn first. The lines of projection are red and solid, but fine. The axes may be either red or blue. Always begin inking by drawing the curves and circles first, oblique lines next, and vertical and horizontal lines last. It is easier to lead a straight line to a curve than the reverse. The dimension lines are red, figures and arrow-points black. The latter must touch the line to which or from which the measurement extends, and should be drawn with a writing-pen. The line between the numerator and denominator of a fraction should always be horizontal.

Fig. 45.—This is the same as Fig. 44, except that the top, as shown in the front view, inclines from a height of four inches on the left to two inches on the right. The lower edge is shown in the side or end elevation as a horizontal line.

Fig. 46.—The front elevation of this figure shows the upper corners of Fig. 44 cut out, one square and the other by an oblique plane. The lower edge of the oblique plane is visible from the end elevation, and is shown as a solid horizontal line. Not so, however, the other end. Any plane on that side which appears as a line in the end elevation is represented by a dotted line. It is frequently desirable to represent in a certain view, lines that cannot be seen from that view, hence they are called invisible lines, and these lines are always dotted.

Fig. 47.—In the front elevation are shown two circles. It should be possible to indicate without the aid of an auxiliary view whether or not these represent circular openings or cylindrical projections. By reference to the top view it is readily seen that the one to the left is a hole and the other a cylindrical projection. Suppose a light thrown upon the object at an angle of 45° from the upper left-hand corner, then in the front view the right-hand and lower edges would be in shadow. These lines, then, are made darker than the others. Likewise with the circles; the opening being shaded toward the top and left-hand side,—the edge casting a shadow,—and the cylinder toward the lower and right-hand side. Now shade the right side and lower lines of each of the three views of the prism.

Plate 7.

Fig. 48.—To draw a rectangular prism having one of its edges oblique to one of the planes of projection.

Whilst in Mechanical Drawing it is intended that objects shall be represented by views seen on a line perpendicular to the planes represented, yet in many cases in Mechanical and Architectural Drawing it is found that there are planes which form various angles other than a right angle with each other, which cannot consequently be shown in their true form and also in their proper relation to each other. As an illustration, take the elevation of a building in which the roof inclines backward and upward.

In Fig. 48 the front elevation is inclined upward toward the right at an angle of 30°. To draw the plan, draw two horizontal lines their true distance apart (one and one-half inches). Project each of the four corners of the front view upward vertically, intersecting and joining the sides of the plan. Then $a'b'$ represents the distance between a and b, $a'c'$ represents the elevation dc, $b'c'$ represents bc.

Whenever a plane inclines obliquely toward the observer it is said to be fore-shortened. Thus, the plane b' c', inclining obliquely toward the observer, is fore-shortened. In the side elevations the true thickness is shown as in the plan, the horizontal edges being obtained by projecting a, b, c, and d across horizontally, thus joining the sides and completing the side view.

Fig. 49.—Here we have the same principle as in the last figure, except that one end of the front elevation inclines forward to the left instead of upward. To obtain this view it is necessary to draw the plan first, inclining upward toward the right at an angle of 30°. Now draw the top and bottom lines of the front view, which are shown their true distance apart (two and one-half inches). Then project each of the four corners of the plan to the top and bottom of the front view.

To find the side view project the top and bottom of the front view across and revolve the four corners from the top view, as shown in the drawing.

Fig. 50.—A different phase of the same principle; the front view inclining upward and forward.

In this case it is necessary to draw the end elevation first, as that is the only view in which the figure is shown in its true form. If the last two figures have been correctly understood, it should be an easy matter to complete this one without further explanation other than that the ends of the plan and front elevation should be drawn with vertical lines, and the horizontal edges projected from the end view.

Fig. 51.—This is exactly the same in form and position as Fig. 49, except that it has an opening through its sides four inches long and one and one-half inches high. In the plan this opening, being invisible, is shown by two dotted lines. The top of the opening is shown in the front and side views by horizontal lines drawn one-half inch below the upper edges of these views. The lower side is drawn one-half inch above the lower edge. Care should be taken that the invisible portions of the two lines are properly shown. The vertical lines in the front and side views of this opening are obtained by projecting from their relative corners in the plan.

Plate 8.

Fig. 52.—To draw a pyramid having an equilateral triangular base.

In this figure a view of the *under* side of the pyramid is drawn first. Begin by drawing a horizontal line three inches long for one side of the lower view.

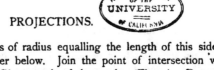

From both ends describe arcs of radius equalling the length of this side (three inches), intersecting each other below. Join the point of intersection with the ends of the horizontal line. Bisect each of the angles (Fig. 2). Draw a horizontal line for the base of front and side views. Also draw indefinitely the axis for front view. On the axis lay off the altitude four and one-half inches above the base. The points in the base of the front view are projected from the lower view as shown in the drawing. The side view is found in the same manner.

Fig. 53.—To draw an equilateral triangular pyramid, one edge of the base resting against the front plane of projection and the axis parallel with the front plane of projection, but inclining toward the right at an angle of 30° with the horizontal.

Draw the axis at the required angle, and upon it construct the base and front elevation as in Fig. 52. Draw the axis in the side view and project through it indefinitely the vertex and points at the base of the front view. Find in the base view the distance from vertex to the nearest point in one of the sides of base, and lay off this distance from axis of side view toward the left. Through this point draw a vertical line, and where it intersects the upper and lower lines projected horizontally from the front view will be two points in the base. Returning again to the base view, find the distance from the vertex to one of the angles. Lay off

3

this distance toward the right from the intersection of axis in side view with the horizontal projection of the middle point in the base of front view. The three base points thus formed are joined with each other and with the vertex, thus completing the side elevation. Where these points revolved to the top view intersect the projections of the same points from the front view, will be found the various points of the top view.

Fig. 54.—To draw a hexagonal pyramid.

Draw a circle two and one-half inches in diameter. With the T-square draw tangents at the top and bottom. Draw four tangents with the 30° triangle. Test the six sides now drawn to see if they are all exactly equal in length. Join opposite angles. Draw the front and side elevations as shown.

Fig. 55.—To draw a hexagonal pyramid with one side of base parallel with the front plane of projection, the axis also being parallel to the front plane of projection, but making an angle of 45° with the horizontal.

Draw the axis at an angle of 45°. Draw the construction circle for base view, and then draw tangents to it by using the T-square and a combination of the 45° and 30° triangles. Join opposite angles. Upon the axis draw the front view. Draw the axis for the side view, and upon it project the vertex and extremities of the base of front view. Upon either side of the axis of side view lay off one-half of the short diameter of hexagon. Through these points draw vertical lines, intersecting them by the projections of the two inner points of base in front view. Join the points in base with each other and with the vertex. Find the top view.

Plate 9.

TRUNCATED PYRAMIDS.

Fig. 56.—To draw a square pyramid having a two-and-one-half inch base and an altitude of four inches, truncated, or cut off, at a point on its axis two inches above the base, by a plane making an angle of 45° with the axis.

Draw the three views *complete*. Cut off the front elevation as shown, with a line drawn at an angle of 45°. Project the intersection of this line with the lines of the sides of this view to the corresponding lines in the other views, and complete the side and top views of the truncated surface.

Fig. 57.—To draw an equilateral triangular pyramid having a base three inches on a side, and an altitude of four inches, truncated by a plane intersecting the axis at an angle of 45°, two inches above the base.

Complete the three views of the pyramid. Through the axis in the front view two inches above the base, draw a line at an angle of 45°. Project the intersection of this line with the lines of intersection of the sides to the corresponding lines in the other view ; thus, o'' on b'' is obtained from o on b', and o' may be revolved from o''. If there were no side view o' might be found by projecting o to s on a', and then projecting vertically s to s' on a. Now, from the vertex in the plan as a centre, revolve s' to o'. The drawing may now be completed.

Fig. 58.—To truncate a hexagonal prism, and show the truncated surface in its true form.

Draw the plan, front and side elevations complete, and proceed with the truncated surface as in the two previous figures. Then draw an auxiliary view as follows : Project the point a' indefinitely toward d'', perpendicular to the line $n\,s\,v$ cutting the front view. Near the upper end of this line draw $o''\,v$ indefinitely and parallel to the trace, $n\,s\,v$ of the cutting plane in front view. Upon the indefinite line $o''\,v$ is found the axis which in the plan is simply a point at o. *A line seen from the end is a point.* Project *o'* to *o''* parallel to *a' d''*. This gives the vertex in the auxiliary view. Now, as the auxiliary view is a view of the plan as seen from a point perpendicular to the line *a o d* inclining upward at an angle of 45° with the ground, the greatest width in the auxiliary view is found on the line of projection from *a'*. On the line *a' d''* from *o'' v* lay off *d''* equal to *o d*. From *o'' v* on *a' d''* lay off *a''* equal to *a*. Project *f'* indefinitely to *e''*, and *b'* indefinitely to *c''*. From *o'' v* on *b' c''* lay off *c''* equal to one-half of *b c*. Lay off *b''* in the same manner. Upon *f' e''* from *o'' v* lay off *e''* equal to one-half of *f e*. Do the same with *f''*. Join *a''*, *b''*, *c''*, *d''*, *e''*, and *f''*. Join each of these points with *o''*. This completes the auxiliary view without showing the cutting plane. The point at which the line of the cutting plane in the front view intersects the line *f' o'* at *n* is now projected perpendicular to the trace of the cutting plane to the line *f'' o''* at *n'*. The line *e' o'* is immediately back of *f' o'*, both appearing as one line. The intersection of this line with the cutting plane is found by continuing the line *n n'* until it intersects the line *e'' o''* at *n''*. Project the intersection of the cutting plane with *a' o'* and *d' o'* at *s* to *s'* on *a'' o''* and *s''* on *d'' o''*. Project *v* on *b' o'* and *c' o'* to *v'* on *b'' o''* and *v''* on *c'' o''*. Join *v'*, *v''*, *s''*, *n''*, *n'*, *s'*, and *v'*, thus showing the true form of the cutting plane, which must of necessity, as heretofore explained, be drawn on a plane perpendicular to *n s v*.

Fig. 59.—To draw an octagonal pyramid truncated by a plane inclining at an angle of 30° with the axis, and intersecting it one and one-half inches above the base.

The main feature in which this figure differs from the last, is, in that the plane cuts through the base, so that in projecting the *lower edge* of the cutting plane from the front view to the other views the line should be drawn to the base and not to the lines of the intersection of the sides.

Plate 10.

Fig. 60.—To draw a hollow pentagonal prism with the ends cut at an angle of 30°, as shown in the drawing.

Draw two concentric circles one and one-quarter inches diameter and two and one-half inches diameter, on which construct the inner and outer pentagons for the end view, from which the dimensions for the front view and plan may be obtained. Lay off five inches for base of front view, and draw upward and inward from these extremities at an angle of 30° with the vertical. In the end view 1 5 is the edge or trace of a plane—the bottom of the prism as it rests upon the ground. The lateral edge of this plane is in the front view directly opposite, and this lateral edge is shown in the end view merely as a point at the extremity, 5. (A line seen from the end is a point.) The other lateral edge 1' as seen from the front view is immediately behind 5', consequently both are represented as one line in the front view. Hence we have 5'=5; 1'=1; 1' being back of 5'. Likewise 2'=2; 3'=3; 4'=4. Number the horizontal lines in the plan the same as those points from which they were revolved in the end view. Project vertically the intersection of the 30° line with 1' to 1''; 2' to 2''; 3' to 3''; 4' to 4''; 5' to 5''. Join 1'' 2'', 2'' 3'', etc. The inner pentagon in the plan is found in the same way by projecting the intersection of the inclined edge 1' 3' with the dotted lines of the inner pentagon in the front view. Repeat for the other end.

Fig. 61.—This is the same as Fig. 60, except that the base of the front view inclines at an angle of 45° with the horizontal, and the plan is shown as viewed obliquely in two different positions. Draw the three views as in Fig. 59, inclining the axis at an angle of 45°, as shown. To draw the right-hand auxiliary view, draw three vertical lines through 1'' 2'' and 3''. The perpendicular distance between 1'' and 2'' equals the perpendicular height of 4 above 5. That of 3'' above 2'' equals the perpendicular height of 3 above 4 2. Project 1' to 1''; 2' to 2''; 3' to 3''; etc. Join 1'', 2'', 3'', 4'', and 5'', thus showing one end of the oblique view minus the opening, which is drawn exactly in the same way. Draw, by the same method, the other end, which of course is invisible. For the upper view, draw horizontal lines through 1''' 2''' and 3''', obtained from 1 2 and 3 as in the previous view. Project 5' and 1' to the base at 5''' and 1'''. Project 4' and 2' to the second line at 4''' and 2''', and lastly project 3' to the top line at 3'''. Join these points. In the previous view this end was visible, but here it is invisible. The other end can now be drawn in the same manner, after which draw the opening as in the last view.

Fig. 62.—To draw the projection of a circular disc when seen at an angle of 30°. Also at 60°.

Draw a line as 7' 19' greater than the diameter of the circle, and at an angle of 60° with the horizontal. From the centre and perpendicular to it draw a line, upon which describe the circle. Divide the circle into any number of equal parts, and project the points to 7' 19' perpendicular to it. From 7' 19' project these points horizontally and vertically as shown. Now, on 1''' 13''' lay off the diameter found at 1 13. On the next horizontal line above, which was projected upward and across from 2, lay off the distance between 2 and 12, one-half on either side of the central vertical line 7''' 19'''. Repeat with the upper lines and also with those below 1''' 13''', the first being taken from 14 24. After these points are found, draw the curve through them, which represents the disc as seen at an angle of 30°. If this is thoroughly understood, it will be an easy matter to draw the upper view, which shows the figure at an angle of 60°, 7″ 19″ showing the apparent distance between 7′ and 19′, and 1″ 13″, as it is perpendicular to 1′ 13′, shows the true diameter.

Plate 11.

THE DEVELOPMENT OF SURFACES.

Fig. 63.—To draw a rectangular prism and show the development of its surfaces.

Draw three views two and one-half inches by one and one-half inches by five inches. Below the front view and some distance from it draw another just like it, two and one-half inches by five inches. Below this last and joined to it, copy the upper view, one and one-half inches by five inches. Continuing downward in the same manner, draw again the front view and then the plan. Opposite the second section and joining it, draw, on either side, views of the end as shown, one and one-half inches by two and one-half inches. This figure when cut out and folded together will form the object represented in mechanical drawing above.

Fig. 64.—To show the development of a cone having an altitude of four and one-half inches and diameter at base of three and one-half inches.

The greatest distance from vertex to base is shown at *a b*.   With *a b* as radius, describe an arc.   Upon this arc lay off the distance *b′ c* equal to the circumference of the base as found in the plan.   Join *b′ a′* and join *c a′*.   At any point on the arc and tangent to it describe a circle equal in diameter to that of the base.   Cut out the figure thus drawn and fold so as to form the complete cone.

Plate 12.

Fig. 65.—To draw the front elevation and plan of a cone five inches high and three and one-half inches diameter at base, cut off by a plane making an angle of 45° with the axis and intersecting it at a point three inches above the base; to draw an auxiliary view showing the true shape of the cut surface, and to draw the development of all the surfaces.

Draw the plan and front view, cutting the latter at an angle of 45° at a point three inches above the base on the axis.   Divide the circle of the base in the plan into any number of equal points, and join them with the vertex at o, as a, o; b, o; c, o; etc. Project the intersection of these lines with the circle to the base of the front view, as a', b', c', etc. It will readily be seen that e' o' is the projection in the front elevation of e o in the plan. Likewise d' o' represents d o. Hence, by projecting vertically the intersection of the trace of the cutting plane in the front view with e' o' to e o, we have the point at which the curved edge of the cutting plane intersects e o in the plan. The intersection of the cutting plane with d' o' is now projected to d o below e o and also to the next line above e o, as in the front view these two lines would be coincident. Project the intersection of the cutting plane with c' o' to c o and the same relative line above e o, and so on with the others. In the absence of a side elevation, the intersection of the cutting plane with a' o' may be projected horizontally to e' o', then upward to e o, and then with o as centre revolved from e o to a o and n o. Join these with a neat curve drawn with the compasses from various centres or with the curved ruler. Draw the line e'' o'' parallel to the trace of the cutting plane in the front view, and project a', b', c', etc., perpendicular to it. Find the curve a'' b'' c'', etc., which is a projection of the circular base upon a plane oblique to it, as in Fig. 62. After drawing the curve, join each of these points with the projection of the vertex at o'', as a'' o'', b'' o'', etc. Project the point of intersection of the cutting plane with a' o' to a'' o'', and also to the same relative line on the other side of e'' o''. Project also the intersection of the cutting plane with b' o' to b'' o'', and so on with the rest. Draw the curve through these points, which, being on a plane perpendicular to the trace as shown in the front view, represents the cutting plane in its true shape and size. Draw the development as in the preceding figure, with the exception of the cut surface. Lay off on the arc from e''' the divisions d''', c''', b''', etc., as many as there are in the plan, and join them with o'''. The true distance from the vertex o' to the point at which the line e' o' is cut is shown in the front view, but not so with d' o', c' o', etc., as these lines incline forward. Project horizontally the inter-section of the cutting plane with d' o', c' o', etc., to e' o', and where these lines of pro-jection intersect e' o' we have the true distance between the vertex o' and the points at which these various lines are intersected by the cutting plane. Transfer these points to the proper lines in the development, as e''' o''', d''' o''', etc., and draw the irregular curve through them. At one end of this curve draw a straight line inclining outward, and perpendicular to this straight line, beginning at the point at which it intersects the irregular curve, draw lines equalling in order, distance, and number the lines projecting the intersection of the cutting plane with e' o', d' o', etc., from the front view to the auxiliary view. The width of the curve on each of these lines in the auxiliary view may readily be copied and transferred, since these two figures are the counterparts of each other. The edge of this surface might have been placed tangent to the irregular curve, but was inclined outward to avoid confusion.

Fig. 66.—To draw the development of a truncated octagonal prism.

Draw the front elevation, plan, and auxiliary view complete, as in Fig. 59. From o''', describe an arc of radius equalling the true distance from the vertex to the base along the edge of any of the inclined planes, and upon this arc lay off eight times the distance $a\,b$. As all of the edges ($a'\,o'$, $b'\,o'$, etc.) of these planes are oblique to the plane of projection, they are not shown in their true length. This may be overcome by drawing an auxiliary view of one of the inclined sides of the figure. Project o' and g' perpendicular to $o'\,g'$. From o'''' draw a line parallel to $o'\,g'$ making $o''''\,n$. From n toward g'''' and f'''' lay off on either side one-half of $f\,g$. Join $o''''\,f''''$ and $o''''\,g''''$. This auxiliary view shows the true shape and size of the plane f, o, g; $f''''\,o''''$ and $g''''\,o''''$ being the true length of the edges. Then with radius $o''''\,f''''$ describe the arc from o'''. Project horizontally the intersection of the cutting plane with $h'\,o'$, $a'\,o'$, etc., to $g'\,o'$. Draw perpendiculars to $g'\,o'$ through these points to $g''''\,o''''$. The points on $g''''\,o''''$ may now be transferred to their respective lines on the development. The truncated surface may be copied literally from the right-hand auxiliary view and placed against any of the edges, as shown in the development. The base may be copied from the plan.

Plate 13.

Fig. 67.—To draw a truncated pentagonal pyramid, and show the development.

Complete the plan and front elevation. Parallel with the trace of the cutting plane in the front view, draw the axis b' o for the auxiliary view. From d'', draw d'' d' perpendicular to b' o. On d'' d', from the intersection of d'' d' with b' o, lay off toward $e f$ the perpendicular distance from b to $e f$. Draw indefinitely the line $e' f'$ parallel with b' o, the points on which are found by projecting from e'' and f'' parallel with d'' d. The points a' and c' are found by projecting from a'' and c'', their distance from b' o being determined by the perpendicular height of c and a above b. On d'' d', from the axis b' o, lay off d' equal to $b d$. Join a', d', c', f', etc. Also join a' b', d' b', c' b', etc. Project the intersection of the cutting plane with the lines in the front view to their respective lines in the auxiliary view. To be enabled to draw the development, it is necessary that we should find the true length of one of the inclined edges of the planes. Draw a vertical line b''' b'''' equal to the altitude of the pyramid. Through b'''' draw a horizontal line, and lay off b'''' a'''' equal to $b a$. Join b''' a'''' the true length of one of the edges. With radius b''' a'''' describe an arc, upon which lay off the five sides of the pentagon and complete the development.

Fig. 68.—To draw a cone five inches high, having a circular base three and one-half inches diameter, cut by a plane making an angle of 45° with the axis, and intersecting the same at a point one and one-half inches above the base.

Draw the plan, front and side elevations, and then also an auxiliary view. If Fig. 65 has been clearly understood, there need be no difficulty apprehended in drawing this one, the methods being similar in both. The appearance of the cut surface differs from the other inasmuch as it intersects the base of the pyramid, from which point in the front view it is projected to the circular base of the plan and to the auxiliary view. The auxiliary view shows the true form of the cut surface, from which it may be copied for the development. Another method for finding the appearance of the cut surface in the plan would be to cut the front view by a series of equidistant horizontal planes, the edges of which would be shown in the plan as circles. Project the intersection of the cutting plane with the traces of these planes to their respective circles in the plan, drawing the curve through the points thus found. As an instance, the base in the front view is shown as a circle in the plan. Project, as has already been done, the intersection of these two lines (the traces of the cutting plane and the base of cone) to the circular base in the plan, and we have the projections of two points of intersection on this trace.

INTERSECTING SOLIDS.

Plate 14.

Fig. 69.—To draw a vertical hexagonal prism having two of its sides parallel with the front plane of projection, intersected by an octagonal prism, having likewise two faces parallel with the front plane of projection, the axes of the two making an angle of 45° with each other.

Draw the plan, front, side, and auxiliary views of the hexagonal prism in the order mentioned, and then draw the inclined axis of the octagonal prism through the front and auxiliary views of the hexagonal prism. Place the needle point of compasses at the intersection of the axes in the auxiliary view, and describe the circle upon which to construct the end of the octagonal prism, the front view of which is now drawn, and lastly, the plan and side view. Let us now consider all but the side view. To find the point at which the plane between g' and h' is intersected by the top of the front right-hand plane of the vertical prism, simply

48

project the intersection of *g h* with *k''* at 1 to *k'* at 1'. Project the intersection of *g h* with the lower edge of the vertical prism at 2 to the vertical line dropped from *k'* at 2'. Join 1' and 2', which is the intersection of the plane *k' m'*, and *g' h'*. The intersection of the plane *i' k'* with *g' h'* is shown at 2' 2'', drawn from 2' parallel with the axis of the inclined prism. The plane *s' i'* is intersected by *g' g'* at 3', found by projecting 3 to *g' g'* at 3'. The intersection 4' 4' is projected to *i'* and *k'* from 4. The intersection of the base *m* with *b' b'* is shown at 5. Join 4' 5. Project 5 to *s i*, and *s o* at 5' and 5'. The intersection of *b''* with *k m* at 1'' is projected to *b' b'* at 1'''. Join 1''' 4'. Project the intersection of *g' g'* with the top of vertical prism to *f''* and *g''*, and join these points, as shown, with 1'' and 1'''',—the projections of 1' on *k m* and *m n*.

To find the intersections in the side view.

The width 1 1 is found at 1'' 1''''. The points 2 and 2 are projected from 2' in the front view, and 2' 2' in the side view are projected from 2'' in the front view. Join 1 2, and 1 2. Project 3' to 3 and 3. The vertical edge *s'* intersects *g' g'* at 6; hence project 6 in the front view to the central line in the side view at 6. Join 2', 3, 6, 3, and 2'. The remaining points of intersection may be readily found by the assistance of the relative numbers. At this point it would be wise for the student to make at least two additional drawings of figures differing from

4

the above in form and position, and show the intersections, for, without such practice in original work, it is difficult to fully impress upon the mind with clearness and permanency the principles involved, though simple they may be.

Plate 15.

Fig. 70.—To draw a square prism and a square pyramid according to the data given in the plate, and show their intersections.

Since the sides of the base of the pyramid incline obliquely toward the front plane, it will be necessary first to draw the plan, and then the front and side elevations. Then, through the axis of the side view, at a point five inches above the base, draw two centre lines forming respectively an angle of 15° with the horizontal and 15° with the vertical, and upon these construct the square end of the prism, thence projecting for the front elevation and plan. The drawing is now completed, excepting the intersections. The points at which the edges of the planes of the pyramid intersect the sides of the prism are all shown in the side view, and may readily be projected to the front view and plan. Project the intersection of gf with $a''o''$ to $a'o'$, and revolve to ao; that of gf with $b''o''$ to $b'o'$ and bo, etc. Project the intersection of he with $a''o''$ to $a'o'$ and ao. Do the same with *b″ o″* and *d″ o″*. It will be seen that *h e* is not intersected by *c″ o″*, but that *c″ o″* cuts *e f.* Project the intersection of *c″ o″* with *e f* to *c′ o′* and *c o*. It will now be seen that a new difficulty arises, inasmuch as the lower corner *e* of the prism intersects the side of the pyramid. To find this point of intersection draw a line through *o″ e* to the base at *s*. This will be the trace of a plane cutting entirely through the pyramid along the edge *e*, that is, the edge between the plane *f e* and *h e*. This plane, then, it is understood, cuts through the base from *s* on the base of the inclined side of the pyramid *o″ b″ c″* to a point immediately back of *s* on the base of the rear inclined side *o″ d″ c″*. The bottom edge, then, of this plane is a line perpendicular to the front plane of projection, and is shown as a point at *s*, and in the plan is shown as a line cutting *c b* at *s″* and *c d* at *s′*. The drawing shows how *s* has been revolved to *s″* and *s′*. A little thought, will show that this cutting plane has three edges, one each where it cuts the base, the side *o″ b″ c″*, and the rear side *o″ d″ c″*. These three edges are shown in the plan. From *s′* to *s″* is the base, *s′* to *o* is the edge along the side *o d c*, and *s″ o* is the edge along the side *o c b*. Hence where *s′ o* intersects *e″* we have *v*, the point at which *e″* pierces the plane *c o d*, and where *s″ o* intersects *e″* we have *v′*, the point at which *e″* pierces the plane *b c o*. Having found the other points, all that remains is to join them, as shown in the plan, and then project *v* and *v′* to the line *e′* in the front view and join the points as found.

Plate 16.

Fig. 71.—To draw a triangular prism and a triangular pyramid of the dimensions given in the plate, the axis of the prism intersecting the axis of the pyramid at an angle of 60°, one side of the prism being parallel with the front plane, and one of the inclined edges making an angle of 15° with the axis of prism, as seen from above.

Draw the plan of the pyramid, and then the front view. Upon the latter draw the inclined axis of the prism, and perpendicular to the axis draw $o\,n$, one side of the prism, n and o being equidistant from the inclined axis. From n, with radius $n\,o$ describe an arc on the axis at m. Join $n\,m$ and $o\,m$. Bisect the angle $m\,o\,n$. Where this line intersects the axis we have the centre of the triangle, through which draw the axis of the auxiliary view of the pyramid, perpendicular to the axis produced of the inclined prism. Find in the plan the perpendicular distance from v'' and w'' to the axis of prism, transferring them on the axis produced of the inclined prism and to the right of its intersection with the auxiliary view of the pyramid. To the left of this point of intersection lay off on the inclined axis the perpendicular distance from s'' to axis of prism in plan, and through the points thus found draw lines parallel with the axis of pyramid. The points s, v, and w are found upon these lines by projecting from s', v', and w', and k on the axis is projected from k'. Join these points, thus completing the auxiliary view of pyramid. Draw the front elevation and plan of prism. The intersections may readily be projected from the auxiliary view to the front view, as, for instance, the intersection of $m\ n$ with $k\ s$ to $k'\ s'$ at 4, etc., and then projected to the plan as 4 to 4', etc. By joining these points as 1, 2, 3, etc., the figure may be completed, but it is not intended that the points of intersection shall be found in this manner. In the previous drawing it was shown how points of intersection were found with the assistance of auxiliary planes, a subject to which we shall now devote considerable attention. Let $a\ b$ represent the trace of a plane cutting the prism on a line with the inclined edge of the pyramid beginning at s''. This cuts the line o'' and n'' at b, and m'' at a. Understand that this imaginary plane cuts entirely through the prism. By projecting b on o'' and n', to b' and b'', on o' and n', and a on m'' to a' on m', and joining a', b', and c', we have the appearance of this plane on the front view. The intersection of the edges $a'\ b'$ and $a'\ b''$ of this plane with $s'\ k'$ at 1 and 4 gives us the points at which the edge $s'\ k'$ pierces the sides $n'\ m'$ and $m'\ o'$ of the prism. In the same manner produce the edge, beginning at w'' to d on m''. From d, to its intersection with o'' and n' at c, we have the trace of a plane cutting the prism through the inclined edge of the pyramid beginning at w''. Project *c* on *o''* and *n''* to *c'* and *c''* on *o'* and *n'*. Project *d* on *m''* to *d'* on *m'*. Then *d' c'* cuts *w' k'* at 2, giving the point at which *w' k'* enters the side *o' m'* of the inclined prism. Join *d'* and *c''*, and its intersection with *w' k'* at 5 shows where *w' k'* intersects the side *m' n'* of the prism. A plane through the inclined edge of which *v''* is the base will not cut *m''*, so it will be necessary to produce *m''* and then drawing *e f* project *e* to *e'* on *o'* and *e''* on *n'* produced. Project *f* to *f'* on *m'* produced. Join *e' f'*, cutting *v' k'* at 3. Join *e'' f'*, cutting *v' k'* at 6. Join 1, 2, and 3, the lines of intersection of the pyramid with the upper side of prism. Join 4, 5, and 6, the lines of intersection on the under side of prism. The points of intersection in the plan are now projected from the front view, 1', 2', and 3' respectively from 1, 2, and 3. The points 4', 5', and 6' are found in the same manner. Prove the accuracy of the work as at 1, 2, 3, 4, etc., by projecting to *s' k'*, etc., from the intersection of *m o* and *m n* with *s k*, etc.

Plate 17.

Fig. 72.—Draw a square prism intersected by a triangular pyramid according to the plate, and show the intersections.

Draw the figures complete. The trace of a plane *a o* intersects *s''* at *o* and *m''* at *a*. Project *a* to *a'* on *m'* and *o* to *o'* on *s'*. Join *a' o'*, intersecting the inclined edge of the pyramid at 1. Points 2 and 3 are found in the same way with *o* and *b* projected to *o'* and *b'* and *o* and *c* projected to *o'* and *c'*. It will be seen that, since the edge *s''* does not pass outside of the pyramid, it must of necessity pierce it. Let *e o d* represent the trace of a plane cutting the pyramid through the edge *s''* of the prism. It is evident that the line *s''* pierces the pyramid somewhere on this plane. Project *e*, the point at which this plane intersects the base of the pyramid, to *e'*, on the base of the front view. Join *e'* with the vertex of the front view, and where this line intersects *s'* we have the point 5, at which the line *s'* pierces one of the inclined sides of the pyramid. This trace gives us also the point 10, where the under edge *n'* of the prism pierces the side of the pyramid. Project *d* to *d'* and join *d'* with the vertex in front view, cutting *s'* at 4 and *n'* at 9, the points at which *s'* and *n'* pierce this inclined side of the pyramid. Join 1, 4, 2, 3, and 5, which shows the intersection of the pyramid with the two upper side of the inclined prism. Join *a' o''* for 6, *b' o''* for 7, and *c' o''* for 8, thus finding thes points at which the edges of the pyramid pierce the under sides of the prism. The points 9 and 10, at which the lower edge of the prism intersects the inclined sides of the pyramid, having already been found, it simply remains to join 6, 9, 7, 8, and 10, thus showing the edges of intersection. These points may now be projected to the plan, and the drawing completed.

Plate 18.

Fig. 73.—To draw two square pyramids, as shown in the figure, with the sides of their bases inclining toward the front plane of projection, and to show their intersections.

The trace of a plane *b a* will show the intersection of one of the edges of the inclined pyramid with the right-hand side of the vertical pyramid. Project *b* to *b′* and project *a* to *a′*. Join *a′ b′*, and at its intersection with the upper edge of the inclined pyramid at 1 we find the point at which the edge of the inclined pyramid pierces the side of the vertical pyramid. Project *d* to *d′* and project *c* to *c′*. Join *d′ c′*, and at its intersection with the edge of the inclined pyramid at 2 we find the point at which that edge pierces the side of the vertical pyramid. Project *h* to *h′* and *g* to *g′*. Join *g′ h′* for 4. Project *f* to *f′* and *e* to *e′*, joining *f′ e′* for 3. Join 1, 2, 3, and 4, which shows the intersection of the inclined pyramid with the side of the vertical pyramid. Project these points to the top view. Project *i* and *k* to *i′* and *k′*, joining *i′ k′* for 5. Project *m* and *d* to *m′* and *d′*, joining *m′* and *d′* for 7. It will be seen that between these two edges of the inclined pyramid there intervenes an edge of the vertical pyramid, hence the edge of the vertical pyramid must pierce the side of the inclined pyramid. We have *b n*, the trace of a plane cutting the side of the inclined pyramid on a line with the edge of the vertical pyramid just mentioned. Project b to the upper edge of the inclined pyramid at b'', and project n to the second edge at n'. Join b'' n', and its intersection with the extreme left-hand edge of the vertical pyramid gives the point 6, at which that edge pierces the upper side of the inclined pyramid. This edge also pierces the front side of the inclined pyramid. The trace of the plane n o intersects the upper edge of the front side of the inclined pyramid at n and the lower edge at o. Project n to n' on the upper edge, and o to o' on the lower edge of this side. Join n' o', and where it intersects the left-hand edge of the vertical pyramid at 8 we find the point at which that edge pierces the front side of the inclined pyramid. The trace of a plane from s to o will help us to find the point at which the under edge of the inclined pyramid pierces the left-hand side of the vertical pyramid. Project s to s' on the base of vertical pyramid, and o to o'' on the extreme left-hand edge of vertical pyramid. Join s' o'', and its intersection with the lower edge of the inclined pyramid at 9 is the point at which that edge pierces the side of the vertical pyramid. Project r and t to r' and t'. Join r' t', and its intersection with the invisible edge of the inclined pyramid at 10 gives the last point to be found. Join 5, 6, 7, 8, 9, and 10. Project the points found to the top view.

Plate 19.

Fig. 74.—To draw a vertical cylinder four inches in diameter and three inches high, intersected by a horizontal cylinder two inches in diameter and one inch long.

Complete the three views according to the drawing. Lay off equidistant points on the circumference of the two-inch cylinder in the side view. Revolve these points to the plan and project them also to the front view. Project the intersection of these lines of projection with the four-inch circle in the plan to the front view, and where these lines of projection in the front view intersect the projections of like numbered points from the end view we have the points through which the curve of intersection of the two cylinders passes. Thus, 7′ is the intersection of the projection of 7 with the projection of 7″, 7″ being revolved from 7. Likewise 8″ is revolved from 8, and 8′ is the intersection of the projection of 8 with the projection of 8″. The other points are found in like manner. Draw through 7′, 8′, 9′, etc., which is the curve of intersection of the two cylinders. To draw the development of the opening in the large cylinder, draw a horizontal line through 10‴ and 4‴. On this line lay off points from 10‴ toward 4‴, as 10″, 9″, 8″, etc., measured on the circumference of the four-inch circle. On 9‴, from the horizontal centre line, lay off the height of 9′ above the centre line in the front view. The perpendicular distance of 11‴ below the horizontal centre in the development equals the perpendicular distance between 11′ and the horizontal centre of the front view. The other points are found in like manner, through which draw the curve of intersection. On the development of the small cylinder, lay off the divisions 10⁗, 9⁗, 8⁗, etc., taken from 10, 9, 8, etc. From 10⁗ upward lay off the distance from 10′ to the right-hand edge of the small cylinder in the front view. From 9⁗ upward lay off the distance from 9′ to the right-hand edge of the horizontal cylinder, etc. Draw through these points, which will form the irregular curved edge of the intersecting end of the small cylinder.

Fig. 75.—To draw a vertical cone intersected by a horizontal cylinder, the axis of the cylinder produced intersecting the axis of the cone at right angles, and to draw the curve of intersection.

Draw the three views of cone and cylinder. The central vertical line in the side view is assumed to be the trace of a plane cutting clear through the cone, and at two points, *a′* and *g′*, on this trace we have points of intersection of the cone and cylinder. The front edge of this plane would trace a line along the extreme right-hand inclined side, *g s*, of the front view; the trace of this plane is shown in the plan at $s'' g$. Project a' and g' to $g s$, and we have two of the points of intersection in the front view, and these points projected upward to $g s''$ in the plan will give the projection of these points on that view. Divide the side view of the cylinder into any number of equidistant points, as a', b', c', etc., and draw traces through these points from the vertex to the base of the cone. Revolve their intersection with the base of cone in this view to the base of cone in the plan and join these points with the vertex, and we have then the traces of these planes on the top view. These may in turn be projected to the front view and the traces drawn, as $h s$, $i s$, etc. The points b', c', d', etc., in the side view may now be projected to the proper lines in the front view and plan, as b' to $b s$ in the front view, and $b s''$ in the plan, and the curve of intersection drawn. On the development of the cylinder lay off the divisions a', b', c', etc., and draw perpendiculars through them. Then a''' is found by copying the distance from the right-hand edge, $u v$, of the cylinder to where the upper edge intersects the trace $g s$. The other measurements are found in like manner. Lay off the development of the cone showing the traces upon its surface. The points a'' and g'' are found on $s g$ where $s g$ is intersected by the projections of a' and g. The other points b'', c'', etc., are found in like manner on $s g$.

Plate 20.

Fig. 76.—To draw a vertical cone having a base four inches in diameter and five and one-half inches high, the axis of which is intersected at right angles by the axis of another cone having a base four inches in diameter and an altitude of eight inches. The axes intersect at a point two and one-half inches above the base of vertical cone, and five and one-quarter inches above the base of horizontal cone. Find the curves of intersection.

For convenience in projecting, draw the plan of the two cones below the front elevation, as shown in the plate. To the left of front view draw one-half of the base of horizontal cone. Produce indefinitely the base of vertical cone, as *a b*, and the horizontal cone, as *c d*. It will be necessary now to pass the auxiliary planes through a line containing the vertices of both cones; hence draw the line *b d*, passing through the vertices *n* and *o*, intersecting also the traces of the planes of the base of each cone at *b* and *d*. The trace *b d* is shown in the plan as *b′ d′*, and in the side elevation *a d″*. Draw any line *b′ e*, which represents the trace of a plane cutting the base of the vertical pyramid at 1 and 2.

Revolve *e* to *e'*. Join *e' d''*, which is tangent to the base of horizontal cone at 3. Project 3 to 3'. Join 3' *o*. Project 1 to 1' and 2 to 2'. Join 1' *n* and 2' *n*, and where these lines intersect 3' *o* gives two points, *k* and *s*, on the curve of intersection. Draw *f b'*. Revolve *f* to *f'*. Join *f' d''*. Project the intersection of *f' d''* with base of horizontal cone at 4 and 5 to 4' and 5', and repeat as with *e b'*. A number of points may be found in this manner. It will readily be seen that the curve, as shown in the plan, is exactly the same on both sides of the plane *d' b'*. Let the student be required to draw the development of these cones, showing the curve of intersection.

Fig. 77.—To draw a rectangular prism, the axis of which inclines to both planes of projection. The front projection of the axis inclines upward at an angle of 60° with the horizontal, and the top inclines forward at an angle of 60°.

Draw *o c*, the front projection of the axis, but of greater length than that of the prism. Draw the top view *o' c'* of this axis. Project *o'* and *c'* perpendicular to *o' c'* toward *o''*. Draw *a' b'* parallel to *o' c'*. From *o''*, toward *o'*, lay off the height of *c* above *a b*, and draw parallel with *o' c'*, cutting the projection of *c'* at *c''*. On *o'' c''*, from *o''*, lay off the true length of the prism at *x*. Draw *v k* through *x* parallel with *a' b'*. The height of *v' k'* above *a b* equals that of *v k* above *a' b'*, and intersects $o c$ at x', thus giving $o x'$, the axis in the front projection. The top of the axis in the front view can also be found by projecting x to $o' c'$, and thence to $o c$. Draw the top view. Project the corners from the top to the front projection. The vertical distance from these corners to $a b$ and $v' k'$ is the same as from $a' b'$ and $v k$.

PLATE 2

PLATE 3

PLATE 4

PLATE 10

PLATE 11

PLATE 14

PLATE 15

PLATE 18

-3.

ze.

PLATE 19